ɔrt

# Consolidating Air Force Maintenance Occupational Specialties

Thomas Light, Daniel M. Romano, Michael Kennedy, Caolionn O'Connell, Sean Bednarz

RAND Project AIR FORCE

For more information on this publication, visit www.rand.org/t/RR1307

Library of Congress Control Number: 2016933045

ISBN: 978-0-8330-9226-7

Published by the RAND Corporation, Santa Monica, Calif.

© Copyright 2016 RAND Corporation

**RAND**® is a registered trademark.

Support RAND
Make a tax-deductible charitable contribution at
www.rand.org/giving/contribute

www.rand.org

# Preface

This report provides select results from a RAND Project AIR FORCE (PAF) project titled *Reducing Operating and Support Costs for the Mobility Air Forces* sponsored by Air Mobility Command (AMC)/A5/8 and AMC/A4. The objective of this project was to identify strategies that can be adopted by the U.S. Air Force to reduce mobility aircraft operating and support (O&S) costs. This report provides an evaluation of the merits of consolidating maintenance occupational specialties for mobility aircraft.

The research was conducted within PAF's Resource Management Program and should be of interest to Air Force personnel concerned with O&S costs and maintenance manpower issues.

## RAND Project AIR FORCE

RAND Project AIR FORCE (PAF), a division of the RAND Corporation, is the U.S. Air Force's federally funded research and development center for studies and analyses. PAF provides the Air Force with independent analyses of policy alternatives affecting the development, employment, combat readiness, and support of current and future air, space, and cyber forces. Research is conducted in four programs: Force Modernization and Employment; Manpower, Personnel, and Training; Resource Management; and Strategy and Doctrine. The research reported here was prepared under contract FA7014-06-C-0001.

Additional information about PAF is available on our website:
www.rand.org/paf

# Contents

# Figures and Tables

## Figures

## Tables

# Summary

In a climate of declining budgets, Air Mobility Command (AMC) is pursuing strategies to reduce aircraft operating and support (O&S) costs without jeopardizing readiness. To assist AMC with its effort, RAND Project AIR FORCE (PAF) considered a number of options targeting unit-level costs. As part of this process, the project team reviewed commercial carrier aircraft maintenance (Mx) approaches and had discussions with subject-matter experts familiar with Air Force aircraft maintenance practices and policies. Based on those interactions, PAF identified consolidation of aircraft maintenance occupational specialties as having the *potential* for reducing personnel requirements and costs. This potential was seen to be worthy of deeper analysis.

Consolidating aircraft maintenance job categories, called *Air Force specialties* (AFSs) is not a new idea for the Air Force. The Air Force underwent a major effort to consolidate the number of maintenance AFSs in the 1980s and 1990s as part of the Rivet Workforce initiative.[1] Newer fighter fleets—the F-22 and F-35—have fewer maintenance AFSs when compared with other legacy fighter platforms. Despite a push toward reducing the number of AFSs, there has been limited quantitative analysis of the resulting impact on manpower requirements, readiness, and cost.

This report presents an analysis of the costs and benefits of consolidating AFSs for mobility aircraft. Specifically, this report seeks to address the following questions:

- What are the current maintainer training requirements, and how might they increase under aircraft maintenance AFS consolidation?
- How will an increase in training requirements from AFS consolidation affect the *availability* of maintainers to perform maintenance?
- How much more *effective* will maintainers be once they are trained on a broader range of tasks under AFS consolidation?
- What is the *cost* of additional required training?
- What is the overall effect of AFS consolidation on cost and readiness?

To address these questions, the impact of AFS consolidation on active duty KC-135 maintenance personnel at MacDill, McConnell, and Fairchild Air Force Bases (AFBs) is modeled.[2]

---

[1] Edward Boyle, Stanley J. Goralski, and Michael D. Meyer, "The Aircraft Maintenance Workforce Now and in the Twenty-First Century," *Air Force Journal of Logistics*, fall 1985, pp. 3–5.

[2] While this analysis is intended to inform KC-46 maintenance initiatives, because data and models specific to the KC-46 are limited, this study focused on the impacts of maintenance AFS consolidation for the KC-135 fleet. KC-135 represents the largest tanker fleet for which there is data on maintainer training requirements and models that can be used to investigate the effect of AFS consolidation on manpower requirements and readiness.

## Analytical Approach

Reducing the number of maintenance AFSs will have positive and negative effects on the maintenance mission. It will require that maintainers spend more time being trained (in technical school and at base performing upgrade and qualification training) and training others, which will reduce a maintainer's *availability* to contribute to the maintenance mission.[3] There are potential benefits of consolidating AFSs that stem from greater *utilization* of maintainers once they receive training on a broader set of tasks. There are also increased training costs, since maintainers must spend more time at technical school learning a broader range of skills. These impacts are depicted in Figure S.1.

**Figure S.1. Impact of AFS Consolidation**

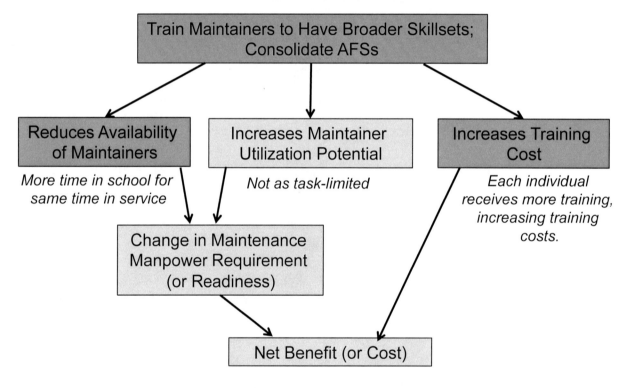

In order to weigh these trade-offs, the current training requirements for Mobility Air Force (MAF) maintainers were reviewed, with a focus on those assigned to support KC-135 maintenance operations. The study team spoke with subject-matter experts to understand how MAF maintainers currently spend their time and developed scenarios of how maintainer training requirements would change following AFS consolidation. Information on attrition rates for

---

[3] Some subject-matter experts have concern that broadening of maintainer task responsibilities could reduce maintainer proficiency or increase the amount of time to conduct certain maintenance tasks. However, others suggested that expanding training and increasing maintainer utilization could increase proficiency and improve maintenance times. This study assumes that consolidating AFSs will not positively or negatively affect the proficiency and speeds with which maintenance tasks are performed.

active-duty maintainers was integrated into this assessment to account for the effect of separations and retention. To analyze the implications of AFS consolidation on maintainer *effectiveness*, the Logistics Composite Model (LCOM) is used. Estimates of the additional costs that Air Education and Training Command (AETC) will incur following AFS consolidation under alternative maintenance force structure assumptions are derived. This information is combined to arrive at overall findings on the manpower requirements, cost, and readiness implications of AFS consolidation. A single consolidation construct that involves reducing the number of KC-135 maintenance AFSs from ten to four is considered, as shown in Figure S.2.

**Figure S.2. Air Force Specialty Consolidation Construct Considered**

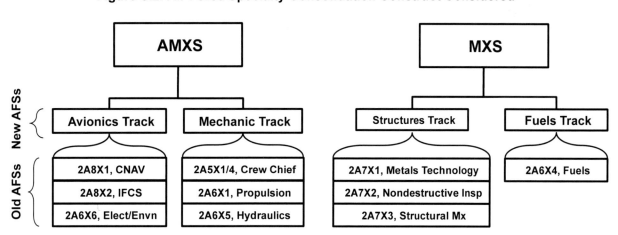

NOTE: CNAV: Communications and Navigation Systems Specialist, IFCS: Instrument and Flight Control Systems Specialist, Elect/Envn: Electrical and Environmental Systems Specialist. This figure represents the primary association of AFSs to the Aircraft Maintenance Squadron (AMXS) and Maintenance Squadron (MXS). However, some AFSs work in both AMXS and MXS.

## Findings

The assessment suggests that combining KC-135 maintenance AFSs will require significant additional time on the part of maintainers to train, which will reduce maintainer availability to perform maintenance activities. In the two scenarios considered, the average reduction in availability varied from 7 to 14 percent after accounting for retention considerations. The reduction in maintainer availability is driven by longer tech-school times and additional qualification and upgrade training requirements. The additional investment in training is assumed to be necessary to maintain proficiency once maintainers are expected to perform a broader range of tasks following AFS consolidation.

The reduced availability of maintainers following consolidation is potentially offset by the improved use and effectiveness of maintainers. Analysis using the LCOM suggests the use and effectiveness benefits of AFS consolidation are likely to be large, and more than offset the reductions in availability caused by additional training requirements. Figure S.3 illustrates the net effect of these offsetting factors at different manning levels. In the figure, the green and blue curves represent the effect of AFS consolidation at different authorized manning levels on sortie generation capability after assuming reductions in maintainer availability of 7 and 14 percent, respectively. These curves should be interpreted relative to the red dot, which represents the current manning authorization level and the associated sortie generation rate.

The findings suggest that consolidating KC-135 maintenance AFSs according to the construct analyzed here would, in the long run, benefit AMC through greater readiness (as measured by sortie generation capability), reduced manpower requirements, or both. If current manpower authorization levels are maintained and fully staffed, KC-135 sortie generation capability would increase by 12 to 14 percent under the AFS consolidation construct analyzed (the blue and green vertical lines in Figure S.3). Alternatively, AMC could reduce its KC-135 maintenance manpower authorization levels by between 17 and 23 percent, without experiencing any loss in sortie generation capability (the blue and green horizontal lines in Figure S.3). A third option involves reducing manpower levels by less than 17 percent, reaping some reduction in manpower costs while also improving sortie generation capability. For example, if AFS consolidation is pursued, AMC could reduce its KC-135 active duty maintenance workforce authorizations by 10 percent (253 positions), increase its sortie generation capability by 7 to 10 percent, and reduce recurring costs by approximately $20 million per year (in fiscal year 14 dollars). Even when considering that maintenance manpower positions are today manned below authorized levels, the same results hold, although the percentage changes are slightly different.

**Figure S.3. Impact of Consolidating Maintenance Air Force Specialties on Sortie Generation Capability and Manpower Requirements**

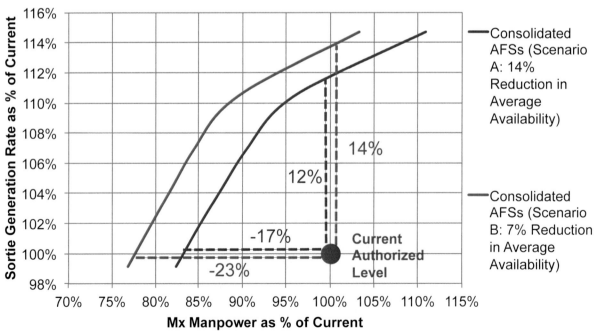

A review of past Air Force experience consolidating maintenance AFSs suggests that implementation details matter and they should be explored more fully before pursuing any consolidation policy. The study team recommends that analysis be conducted in the following additional areas to help the Air Force further evaluate the merits of maintenance AFS consolidation for legacy and new fleets such as the KC-46:

- Investigate the impact on Guard and Reserve maintainers.
- Identify implementation challenges and develop strategies for reducing implementation issues.
- Evaluate the effects on other fleets.
- Refine estimates of training requirements associated with technical school and upgrade and qualification training.
- Evaluate the impacts on accession and promotion requirements.
- Evaluate the impacts on deployed footprint and combat resiliency.

# Acknowledgments

The study team appreciates the sponsorship of this research by Maj Gen Michael Stough (AMC/A5/8) and Maj Gen Warren Berry (AMC/A4).[4] Maj Michael Germany (AMC/A5/A8X) provided excellent support as the primary point of contact for this study.

In addition, many people provided helpful feedback during this research. They include Lt Gen Judy Fedder (AF/A4/7), Maj Gen John Cooper (ACC/A4), Brig Gen Katheryn Johnson, (AF/A4/7), Daniel Fri (AMC/A4), David Merrill (AMC/A9), and Col Walter Isenhour (AMC/A4). The study team is especially thankful for the input provided to us on this effort by a number of subject-matter experts, including Lt Col Elizabeth Clay (AMC/A4M), Sean Spano (AMC/A4M), CMSgt Todd Krulcik (AMC/A4M), James Booth (AMC/A4M), Col Tony Pounds (AETC A4/7D), Bruce Brune (AMC/A1M), Shenita Clay (3rd MRS), and a number of people at United Airlines, especially John Kelly and Bill Gilbert.

A number of colleagues at RAND, including John Drew, Laura Baldwin, Al Robbert, and Ray Conley, provided support during this effort. Laura Baldwin, Patrick Mills, and Jim Powers provided helpful comments on an early draft of this report. John Crown and Brent Thomas formally reviewed this report and provided numerous helpful comments and suggestions. Brent Thomas and Muharrem Mane provided guidance and help with the study team's implementation of LCOM. Col Geof Nieboer (AF/A4/7P) participated in the initial stage of this research and provided excellent input while visiting RAND as an Air Force Fellow. Megan McKeever and Donna Mead provided administrative support.

Acknowledgment of these individuals does not imply their endorsement of the views expressed in this report.

---

[4] All offices and ranks are current as of the time of the research.

# Abbreviations

| | |
|---|---|
| A&P | Airframe and Powerplant |
| AARA | Aircraft Airworthiness Release Authority |
| AETC | Air Education and Training Command |
| AFB | Air Force Base |
| AFS | Air Force specialty |
| AFSC | Air Force specialty code |
| ALS | Airman Leadership School |
| AMC | Air Mobility Command |
| AMXS | Aircraft Maintenance Squadron |
| BMT | Basic Military Training |
| CFETP | Career Field and Education Training Plan |
| CNAV | Communications and Navigation |
| CUT | cross-utilization training |
| Elect/Envn | Electrical and Environmental Systems Specialist |
| ELEN | Electro-Environmental |
| FAA | Federal Aviation Administration |
| FY | fiscal year |
| IFCS | Instrument and Flight Control Systems |
| LCOM | Logistics Composite Model |
| MAF | Mobility Air Force |
| MDS | mission design series |
| MOB | main operating base |
| MQTP | Maintenance Qualification Training Program |
| MSgt | Master Sergeant |
| Mx | maintenance |
| MXS | Maintenance Squadron |
| NCO | noncommissioned officer |
| NCOA | Noncommissioned Officer Academy |
| NOC | Network Operation Center |
| O&S | operating and support |
| PAA | primary aircraft authorizations |
| PAF | Project AIR FORCE |
| PME | professional military education |
| SMART | Simulated Maintenance and Requirements Tool |
| SrA | Senior Airmen |

| | |
|---|---|
| TACC | Tanker Airlift Control Center |
| TOMC | Technical Operational Maintenance Control |
| UA | United Airlines |
| YOS | year of service |

# 1. Introduction

In a climate of declining budgets, Air Mobility Command (AMC) is pursuing strategies to reduce aircraft operating and support (O&S) costs without jeopardizing readiness. To assist AMC with its effort, RAND Project AIR FORCE (PAF) considered a number of options targeting unit-level costs. The project team reviewed commercial carrier aircraft maintenance (Mx) approaches and had discussions with experts on Air Force aircraft maintenance practices and policies.[5] Based on those interactions, PAF identified consolidation of aircraft maintenance occupational specialties, or job categories, as having the potential for reducing personnel requirements and costs.

PAF analyzed the manpower and readiness implications of consolidating the number of maintenance job categories, called *Air Force specialties* (AFSs), which support the KC-135 to inform KC-46 maintenance initiatives. Because data and models specific to the KC-46 are limited, this study focused on the impacts of maintenance AFS consolidation for the KC-135 fleet. KC-135 represents the largest tanker fleet for which there is data on maintainer training requirements and models that can be used to investigate the effect of AFS consolidation on manpower requirements and readiness.

AFS consolidation is not a new idea for the Air Force. Layne et al. note that, between 1984 and 1997, the Air Force reduced the number of occupational categories it employed by 10 percent.[6] The Air Force underwent a major effort to consolidate the number of maintenance AFSs in the 1980s and 1990s as part of the Rivet Workforce initiative.[7] Newer fighter fleets—the F-22 and F-35—have fewer maintenance AFSs when compared with other legacy fighter platforms. Despite a push toward reducing the number of AFSs, there has been limited quantitative analysis of the resulting impact on manpower requirements, readiness, and cost.[8]

This report quantifies the costs and benefits of consolidating maintenance AFSs for the KC-135 fleet. Specifically, this report seeks to address the following questions:

- What are the current training requirements, and how might they increase under aircraft maintenance AFS consolidation?
- How will an increase in training requirements from AFS consolidation affect the *availability* of maintainers to perform maintenance?

---

[5] Appendix A provides a summary of differences observed between maintenance practices performed at United Airlines and those used by the Air Force to maintain mobility aircraft.

[6] Mary E. Layne, Scott Naftel, Harry J. Thie, and Jennifer H. Kawata, *Military Occupational Specialties: Change and Consolidation*, Santa Monica, Calif.: RAND Corporation, MR-977-OSD, 2001.

[7] Boyle, Goralski, and Meyer, 1985.

[8] An important exception to this is Layne et al., 2001, which evaluated the readiness impacts of consolidating a broad number of Army and Marine occupational specialties. They find no evidence of reduced readiness following consolidation.

- How much more *effective* will a maintainer be once they are trained on a broader range of tasks under AFS consolidation?
- What is the *cost* of additional required training?
- What is the *overall effect* of AFS consolidation on cost and readiness?

The analysis presented here considers the impact of AFS consolidation on active-duty KC-135 maintenance personnel at MacDill, McConnell, and Fairchild Air Force Bases (AFBs).[9] Assessment of the impact of AFS consolidation on other mobility fleets and Guard and Reserve personnel was beyond the scope of this study.

## Occupational Specialty Consolidation Construct

This analysis considers consolidation of aircraft maintenance AFSs that are typically employed on the flight line and in back shops.[10] A single construct for combining ten Air Force specialty codes (AFSCs)[11] into four is analyzed, which is shown in Figure 1.1. Many other reasonable approaches for consolidating AFSs exist and are worthy of consideration. We arrived at the consolidation construct shown in Figure 1.1 after (1) reviewing the way in which commercial airlines organize their maintenance workforce, (2) talking with Air Force subject-matter experts, and (3) reviewing past proposals for AFS consolidation developed by others.[12] The consolidation

---

[9] PAF has recently studied the cost and readiness implications of other forms of consolidation. For example, McGarvey et al. (Ronald G. McGarvey, Manuel Carrillo, Douglas C. Cato, John G. Drew, Thomas Lang, Kristin F. Lynch, Amy L. Maletic, Hugh G. Massey, James M. Masters, Raymond A. Pyles, Ricardo Sanchez, Jerry M. Sollinger, Brent Thomas, Robert S. Tripp, and Ben D. Van Roo, *Analysis of the Air Force Logistics Enterprise: Evaluation of Global Repair Network Options for Supporting the F-16 and KC-135*, Santa Monica, Calif.: RAND Corporation, MG-872-AF, 2009), Tripp et al. (Robert S. Tripp, Ronald G. McGarvey, Ben D. Van Roo, James M. Masters and Jerry M. Sollinger. *A Repair Network Concept for Air Force Maintenance: Conclusions from Analysis of C-130, F-16, and KC-135 Fleets*, Santa Monica, Calif.: RAND Corporation, MG-919-AF, 2010), and Van Roo et al. (Ben D. Van Roo, Manuel Carrillo, John G. Drew, Thomas Lang, Amy L. Maletic, Hugh G. Massey, James M. Masters, Ronald G. McGarvey, Jerry M. Sollinger, Brent Thomas, and Robert S. Tripp, *Analysis of the Air Force Logistics Enterprise: Evaluation of Global Repair Network Options for Supporting the C-130.* Santa Monica, Calif.: RAND Corporation, TR-813-AF, 2011) considered geographically centralizing certain heavy maintenance and inspection activities for mobility aircraft including the KC-135. In another line of research, McGarvey et al. (Ronald G. McGarvey, James H. Bigelow, Gary J. Briggs, Peter Buryk, Raymond E. Conley, John G. Drew, Perry Shameem Firoz, Julie Kim, Lance Menthe, S. Craig Moore, William W. Taylor, and William A. Williams, *Assessment of Beddown Alternatives for the F-35*, Santa Monica, Calif.: RAND Corporation, RR-124-AF, 2013) considered savings that might arise from adopting a more consolidated basing construct (e.g. operating large squadrons out of fewer bases) for the F-35. This analysis considers the potential cost and readiness impacts that might result from a third form of consolidation, namely the consolidation of career fields.

[10] Cross-utilization training (CUT) is a related practice. Under CUT, a maintainer receives training that enables he or she to perform tasks that are not in their primary AFS. CUT is sometimes employed to address manning shortfalls. For a discussion of the use of CUT by the Army for helicopter field maintenance, see William G. Wild, Bruce R. Orvis, Rebecca Mazel, Iva Maclennan, and R. D. Bender, *Design of Field-Based Crosstraining Programs and Implications for Readiness*, Santa Monica, Calif.: RAND Corporation, R-4242-A, 1993.

[11] Each AFS is assigned an AFSC. In practice, AFS and AFSC are often used interchangeably.

[12] In 2008, Booz Allen Hamilton completed a study for AF/A4MM that described an approach for development and implementation of a strategic roadmap for integrating planned maintenance workforce initiatives with the proposed training business model and organizational structure developed as part of the Training Enterprise 2010 Concept of

construct considered here combines career fields with similar tasks and training requirements. Depending on the extent by which there is overlap in tasks and training, consolidation of AFSs will have differing impacts on the amount of extra training and time required to advance in skill level.

Under the construct shown in Figure 1.1, the six primary Organizational Level Maintenance/sortie generation specialties in the Aircraft Maintenance Squadron (AMXS) are combined into an avionics and a mechanics track, while the four primary intermediate-level maintenance/backshop AFSCs in the Maintenance Squadron (MXS) are combined into a structures and fuel track.[13] The choice of an avionics and mechanics track on the flight line is similar to the flight line AFS structure for the F-35, which includes a crew chief (2A3X7)[14] and avionics (2A3X5) track. Furthermore, the three AFSCs making up the avionics track are similar to the avionics track observed at United Airlines (UA).

**Figure 1.1. Air Force Specialty Consolidation Construct Considered**

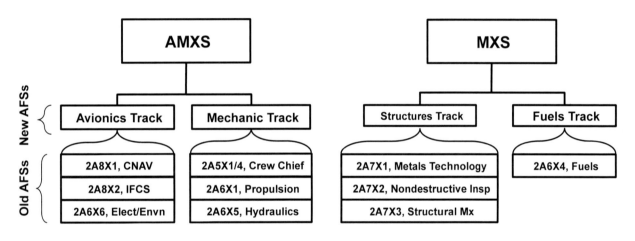

NOTE: CNAV: Communications and Navigation Systems Specialist, IFCS: Instrument and Flight Control Systems Specialist, Elect/Envn: Electrical and Environmental Systems Specialist. This figure represents the primary association of AFSCs to AMXS and MXS. However, some AFSs work in both AMXS and MXS.

In the backshop, metals technology (2A7X1), nondestructive inspection (2A7X2), and structural maintenance (2A7X3) are combined into a structures track because of the similarity in maintenance tasks and training requirements. This consolidation is similar to the structures

---

Operations. (See Matthew McMahan, *Maintenance Force Development Study: Phase II Final Report*, Booz Allen Hamilton, April 2, 2007; and Matthew McMahan, *Maintenance Force Development Study: Final Report*, Booz Allen Hamilton, June 17, 2008.) This study suggested developing a multiskilled flight line maintainer consisting of a mechanical and technical (aka Mech/Tech) track but did not assess the overall impact on readiness or manpower requirements.

[13] It is understood AMXS AFS are assigned to MXS in small numbers and will still be required under a consolidated construct.

[14] Note that the AFSC for the crew chief track for the F-35 is different from the crew chief AFSC for cargo/tanker aircraft.

occupation observed at UA. The fuels track (2A6X4) was left alone because it did not logically fit into the structures track. In this analysis, it is assumed that there is no change to the way in which maintainers in the fuel track are trained.

The consolidated AFS structure analyzed in this report was vetted with subject-matter experts at AMC and Air Education and Training Command (AETC) and deemed sensible for analysis. Nevertheless, the study team readily acknowledges that other consolidation constructs could be justified and may offer additional advantages. It was beyond the scope of this study to explore a wider range of possible consolidation constructs.

## Analytical Approach

Reducing the number of maintenance AFSs will have positive and negative effects on the maintenance mission. It will require that maintainers spend more time being trained (in technical school and at base performing upgrade and qualification training) and training others, which will reduce a maintainer's *availability* to contribute to the maintenance mission. There is also the potential for increased training costs, particularly by AETC, since maintainers must spend more time at technical school learning a broader range of skills. There are, however, potential benefits of consolidating AFSs. In particular, there is the potential for greater use of maintainers who are less task limited once they receive training on a broader set of tasks. These impacts are depicted in Figure 1.2.[15] The red boxes indicate costs associated with the consolidation, and the green box indicates a benefit. The quantitative results from the red and green boxes will determine whether the gray boxes should be considered a net benefit or a net cost.

---

[15] In the course of conducting this research, other potential costs and benefits of AFS consolidation were identified that are not explicitly considered here. For example, one-time investments or disruptions that will be experienced when transitioning to a more consolidated set of AFSs are not considered. Similarly, the potential for AFS consolidation to result in a smaller deployed footprint has been identified as a possible benefit but was not thoroughly analyzed by the study team. It has also been noted that workers experience greater job satisfaction from less restrictive job categories that result from consolidation because it is easier to find jobs that align with their personal preferences (Layne et al., 2001). Finally, consolidating career fields has the potential to reduce maintainer proficiency, as airmen are expected to perform a wider range of tasks. Conversely, consolidation of career fields would likely lead to expanded training and has the potential to increase maintainer utilization, which could increase proficiency. This study assumes that consolidating AFSs will not positively or negatively impact the proficiency and speeds at which maintenance tasks are performed.

**Figure 1.2. Impact of Air Force Specialty Consolidation**

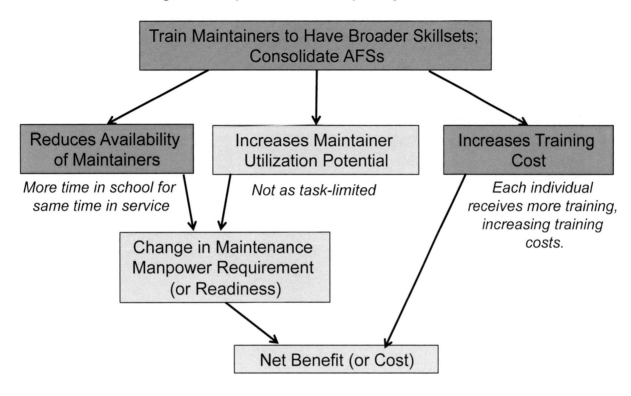

This analysis seeks to quantify the costs and benefits shown in Figure 1.2 to inform decisionmakers at AMC on the merits of AFS consolidation. For availability of maintainers, training requirements for Mobility Air Force (MAF) maintainers were reviewed, with a focus on those assigned to support KC-135 maintenance operations. The research team spoke with subject-matter experts to understand how MAF maintainers currently spend their time. Input from subject-matter experts was used to develop two scenarios that describe how maintainer training requirements will change following AFS consolidation. Information on attrition rates for active-duty maintainers was integrated into this analysis to account for the effect of separations.

To analyze the implications of AFS consolidation on maintainer *utilization* and *effectiveness*, the Logistics Composite Model (LCOM) and the official KC-135 LCOM input database is used. LCOM is used by the Air Force to derive wing/squadron maintenance manpower requirements and can also be adapted to look at how policy changes such as AFS consolidation affects measures of readiness, such as sortie generation rates under high sortie demand schedules. Within the KC-135 LCOM, the manpower levels are varied under the existing AFSC construct and the consolidated construct to observe how sortie generation rates vary in a high sortie–demand environment. This information is combined with estimates of the reduced availability of maintainers to arrive at overall findings on manpower requirement and readiness implications of AFS consolidation.

Estimates of the additional costs that AETC will incur following AFS consolidation under alternative assumptions about the size of the workforce are derived. This assessment used data on the current AFS course times and costs, provided to the project team by AETC.

## Structure of This Report

In addition to this introductory chapter, Chapter Two presents the analysis of the impact of AFS consolidation on maintainer availability. Chapter Three integrates the findings from Chapter Two, with analysis conducted using LCOM to determine how AFS consolidation affects maintainer effectiveness. Based on this assessment, it is shown how AFS consolidation affects KC-135 sortie generation capability at different manning levels. Chapter Four examines training costs, quantifies the long-run cost-saving potential of AFS consolidation, and concludes with a discussion of caveats and future analysis that should be conducted.

# 2. Impact of Air Force Specialty Consolidation on Maintainer Availability

A critical step in understanding the costs of AFS consolidation is to quantify the additional training burden that it will place on maintainers. This analysis seeks to quantify the amount of time maintainers are available to perform maintenance under the current and consolidated AFS construct, as shown in Figure 1.1.

To accomplish this, the study team first characterized how time is spent on different activities over a 16-year career for each existing and consolidated AFS.[16] This provides a baseline from which to consider a maintainer's availability to perform maintenance activities. The estimates are adjusted to reflect maintainers' availability to perform maintenance based on retention rates implied from the Air Force's current years-of-service profile for active-duty maintainers. The assumptions and findings on maintainer-availability impacts are discussed in this chapter.

## Estimating Maintainer Availability Over a 16-Year Career

Aircraft-maintenance personnel spend a significant portion of their time engaged in training and non-maintenance activities (Drew et al., 2008). In this section, data and input obtained from subject-matter experts paint a picture of how maintainers spend their time over the course of their career, under the current and consolidated AFS structure (as depicted in Figure 1.1 in the previous chapter).

Because there is uncertainty in how AFS consolidation will be implemented and its effect on training requirements, two scenarios were developed that span a range of perspectives and assumptions:[17]

- *Scenario A* is intentionally meant to be conservative in its characterization of the training requirements imposed on maintainers from consolidating AFSs. That is, this scenario is meant to represent an upper bound on the training requirements that maintainers are likely to face if AFSs are combined as outlined in the previous chapter. This scenario does not eliminate any training duplication that will likely exist in technical school and Career Field and Education Training Plans (CFETPs) under a consolidated construct.

---

[16] After 16 years, the typical technical career of an Air Force maintainer ends, as they are promoted to the rank of Master Sergeant (MSgt). Once attaining the rank of MSgt, the maintainer generally moves from being a first-line technical expert and supervisor to an operational leader and manager. At this point, the maintainer has moved away from hands-on aircraft maintenance (i.e., so-called wrench turning). That is, the Air Force relies on maintainers within their first 16-year period of service to provide the hands-on portion of aircraft maintenance productivity.

[17] While the training requirements are varied in both these scenarios, in the next chapter, maintainers are assumed to be equally competent under both training scenarios.

- *Scenario B* is more consistent with the expectations of subject-matter experts in terms of how AFS consolidation will actually be implemented (based on conversations with the study team). This scenario assumes that the first four weeks of training for all AFSs are equivalent and cover fundamentals such as flight-line safety and aircraft launch and recovery procedures. It also assumes less upgrade and qualification training is required at an airman's duty station relative to Scenario A. These training-time reductions are assumed to be obtainable without reducing training depth and by eliminating duplicative training requirements that are unnecessary following AFS consolidation.

Below, training and other assumptions used in the analysis of maintainer *availability* are discussed. This includes assumptions about time spent on professional military education (PME), technical school, leave, out-of-hide duties, supervision and management, and being trained and training others. See Figure 2.1 for a representation of the timing and magnitude of factors affecting the availability of a crew chief to perform maintenance activities under the current system.

## Professional Military Education

Every enlisted service member starts his or her military career with Basic Military Training (BMT), which lasts approximately 2.3 months (70 days).[18] After about four years of service, maintainers will spend approximately 1.2 months (35 days) at Airman Leadership School (ALS). Following ALS, maintainers can take on supervisory and management roles. At 12 years of service, maintainers will typically spend 1.4 months (42 days) at Noncommissioned Officer Academy (NCOA).[19] The three PME events are depicted in purple in Figure 2.1. The timing and length of PME is assumed to remain the same after AFS consolidation.

## Technical School

After completing BMT, airmen attend technical school, where they learn maintenance skills required to work in their assigned area of specialization. The amount of time (and number of locations where training is performed) varies by AFS. Table 2.1. shows the current *pipeline times*[20] for existing AFSs and those assumed after consolidation. Time spent in technical school is depicted in pink in the example provided in Figure 2.1.

---

[18] With travel included, BMT is slightly greater at 2.6 months, as reflected in Table 2.2.

[19] In 2014, the NCOA course was changed to include a 12-month-long correspondence course designed to provide foundational concepts to NCOs, prior to the four week in-residence Intermediate Learning Experience. All airmen with seven to 12 years in service (regardless of rank) are eligible to take the new NCOA course, while airmen with more than 12 years in service are eligible to take the computer-based training portion only (Mozer O. Da Cunha, "NCOA Changes Curriculum and Eligibility Requirements," *Inside Barksdale AFB*, November 4, 2014, updated November 5, 2014). These changes make it possible for airmen to attend NCOA earlier in their career than assumed in our analysis.

[20] Pipeline times vary by AFS and include time allocated to transit to and from technical school and time spent in the classroom. Pipeline times include weekends and other unproductive time.

Scenario A adds together the time spent in the classroom for each combined AFSC. Transit time is then added for training to occur at either one or two locations. In Scenario B, the first four weeks of classroom time are assumed to overlap for combined AFSs and, therefore, need not be repeated.[21]

As an example, a KC-135 crew chief currently spends 2.3 months (70 days) in the technical school "pipeline." This includes 60 days of classroom time and ten days in transit. Once the crew chief AFSC is folded into the mechanics track, under Scenario A, a mechanic will spend an additional 89 days in technical school learning propulsion and hydraulics skills, for a total of 159 pipeline days[22] or 5.3 months.[23] Under Scenario B, it is assumed that a mechanic will spend an additional 29 days relative to a crew chief currently at technical school learning propulsion and hydraulics skills, for a total time at technical school of 99 days or 3.3 months.

**Table 2.1. Technical School Pipeline Months Before and After Air Force Specialty Consolidation**

| | | | Currently (months) | After Consolidation (months) | |
| | | | | Scenario A | Scenario B |
|---|---|---|---|---|---|
| Avionics | 2A8X1G | CNAV | 4.2 | 10.4 | 8.4 |
| | 2A8X2G | IFCS | 3.5 | | |
| | 2A6X6 | ELEN | 3.2 | | |
| Mechanic | 2A5X1/4A | Crew Chief | 2.3 | 5.3 | 3.3 |
| | 2A6X1C | Propulsion | 1.5 | | |
| | 2A6X5 | Hydraulics | 1.8 | | |
| Structures | 2A7X1 | Metals | 2.4 | 6.8 | 4.8 |
| | 2A7X2 | NDI | 1.8 | | |
| | 2A7X3 | Structures | 2.7 | | |
| Fuels | 2A6X4 | Fuels | 1.3 | 1.3 | 1.3 |

NOTE: Pipeline times were provided to RAND by AETC in July of 2014. Pipeline days following AFS consolidation were estimated by RAND using the methodology described above.

---

[21] Note that the assumption that technical school times extend following consolidation runs counter to Layne et al., 2001, which claims that "[a]s occupational specialties are combined, schoolhouse training may become shorter and more general, while more-specific tasks are learned through on-the-job training." Wild et al., 1993, provide some empirical evidence for this. They found that the Army reduced technical school training times when it consolidated helicopter maintenance units. Nevertheless, the assumption that technical school training times will need to increase to maintain adequate depth of knowledge among maintainers is maintained.

[22] 159 days = 149 days in the classroom + ten days for transit.

[23] Currently KC-135 maintainers specializing in propulsion (2A6X1C) and hydraulics (2A6X5) spend 40 and 49 days in the classroom, respectively. Five days are allocated to transit for both these AFSCs.

## Leave

Currently, airman receive one month of leave per year. Leave is represented in green in the example shown in Figure 2.1. It is assumed that the amount of leave provided does not change after AFS consolidation is implemented.

## Out-of-Hide Duties

Maintainers must perform so-called out-of-hide tasks as part of their employment responsibilities. These include, for example, responsibilities associated with being the squadron resources manager, squadron small computer manager, dormitory manager, squadron safety noncommissioned officer (NCO), and squadron mobility NCO. Time spent on out-of-hide duties can vary significantly across individuals and over one's military career. Out-of-hide duties are assumed to reduce an active duty maintainer's availability to perform maintenance by 0.7 months per year, on average.[24] Out-of-hide duties are shown in yellow in the crew chief example depicted in Figure 2.1.

**Figure 2.1. Share of Time Spent on Different Activities for a Crew Chief Under Current System**

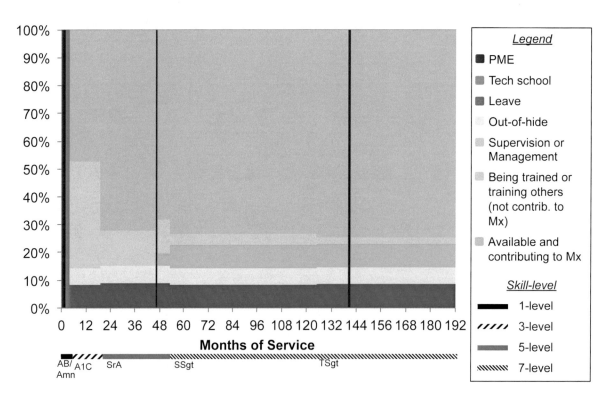

---

[24] This is in line with the share of authorized maintenance positions associated with out-of-hide positions reported in John Drew, Kristin F. Lynch, James M. Masters, Robert S. Tripp, and Charles Robert Roll, Jr., *Options for Meeting the Maintenance Demands of Active Associate Flying Units*, Santa Monica, Calif.: RAND Corporation, MG-611-AF, 2008.

## Supervision or Management

Senior Airmen (SrA) and NCOs spend a portion of their time performing supervisory or management functions, such as mentoring, providing subordinate feedback, writing enlisted performance reports, and managing subordinate training and career development. For SrA who have graduated from ALS, 0.5 months per year are allocated to supervision and management duties; for NCOs, one month per year is allocated for supervision and management activities. Supervisory or management roles are shown in blue in the example depicted in Figure 2.1.

## Being Trained and Training Others

Once arriving at their duty stations, maintainers will spend a significant portion of their time on the Maintenance Qualification Training Program (MQTP) and on-the-job training. As part of this training, airmen spend some of their time working on an aircraft, learning or using new skills required by their specialty field. As maintainers advance in training, they move up in skill level, from 3- to 5-levels to 7-level.

Under the existing training construct, it is assumed that maintainers spend approximately 15 months as a 3-level and 34 months as a 5-level before transitioning to 7-level status. It is assumed that 75 percent of a maintainer's time performing maintenance as a 3-level is in a training capacity (being trained), and this falls to 25 percent as a 5-level maintainer, and 5 percent once a maintainer reaches 7-level status. Under the consolidated AFS scenarios, the amount of time that maintainers spend as a 3-level is extended, and the amount of time spent as a 5-level is reduced. The fraction of time maintainers spent training others as a 5- and 7-level maintainer is also increased, as outlined in Table 2.2.[25] It is assumed that 40 percent of the time that a maintainer spends being trained contributes to the maintenance mission. This assumption is consistent with previous RAND research.[26]

Once maintainers become a 5-level, they start to train other maintainers. Under the current system, it is assumed that maintainers at the 5- and 7-level spend on average 1.8 months per year training others. To accommodate the additional training requirements of incoming airman after AFS consolidation is implemented, the amount of time 5- and 7-level maintainers spend training others is increased to 3.6 and 2.7 months per year in Scenarios A and B, respectively.[27] It is assumed that maintainers are 85 percent productive at contributing to the maintenance mission when they are training others.[28] Time that maintainers spend doing independent maintenance

---

[25] The assumption made regarding extending the amount of 3-level time and reducing the amount of 5-level time after AFS consolidation effectively increases the amount of time that an airman will spend training before they become a 7-level, since training is more intense as a 3-level than as a 5-level.

[26] Drew et al., 2008.

[27] The assumptions on training times under the existing and consolidated AFS construct were developed with input from subject-matter experts.

[28] The assumptions on training and trainer productivity levels were derived by Drew et al., 2008, from Steven A. Oliver, *Cost and Valuation of Air Force Aircraft Maintenance Personnel Study*, Maxwell AFB, Gunter Annex, Ala:

11

(not being trained or training others) is assumed to be 100 percent effective. The example shown in Figure 2.1 depicts in orange the amount of time that a crew chief is being trained or training others and not contributing the maintenance mission under the current system.

**Table 2.2. Assumptions About Time at Skill Level and Training**

| | Currently | After Consolidation | |
| --- | --- | --- | --- |
| | | **Scenario A** | **Scenario B** |
| **Time at Skill Level** | | | |
| 1-level | | | |
| PME | 2.6 months | 2.6 months | 2.6 months |
| Tech School | See Table 1 | See Table 1 | See Table 1 |
| 3-level | 15 months | 27 months | 24 months |
| 5-level | 34 months | 27 months | 24 months |
| 7-level | Time remaining to 16 years of service | Time remaining to 16 years of service | Time remaining to 16 years of service |
| **Percent of Time Available to Perform Maintenance That Is Spent Training** | | | |
| 1-level | NA | NA | NA |
| 3-level | 75% | 75% | 75% |
| 5-level | 25% | 60% | 42.5% |
| 7-level | 5% | 10% | 7.5% |

## Available and Contributing to Maintenance

Using the assumptions outlined above, the amount of time that maintainers are available and potentially contributing to the maintenance mission for each of the existing and consolidated maintenance AFSCs is calculated.[29] In Figure 2.1, this area is depicted in gray.

Figure 2.2 allows one to compare how a crew chief under the existing system and a mechanic under the consolidated Scenario A would spend his or her time over the course of a 16-year career. The amount of time that the maintainer is projected to be available and potentially contributing to the aircraft maintenance mission changes from 132 equivalent months for a crew

---

Air Force Logistics Management Agency, August 2001; Mark J. Albrecht, *Labor Substitution in the Military Environment: Implications for Enlisted Force Management*, Santa Monica, Calif.: RAND Corporation, R-2330-MRAL, 1979, and Carl J. Dahlman, Robert Kerchner, and David E. Thaler, *Setting Requirements for Maintenance Manpower in the U.S. Air Force*, Santa Monica, Calif.: RAND Corporation, MR-1436-AF, 2002.

[29] The calculation of time "available and contributing to maintenance" incorporates the assumptions on effectiveness outlined in the previous section. For example, if a maintainer spends one hour being trained, they contribute 24 minutes (= 60 minutes x 40 percent effective) worth of equivalent time to the maintenance mission.

chief under the current system to 120 months for a mechanic under Scenario A—a reduction in availability to perform maintenance activities of approximately 9 percent.

**Figure 2.2. Comparison of Time Allocation for a Crew Chief Under the Current System and a Mechanic Under Consolidation Scenario A**

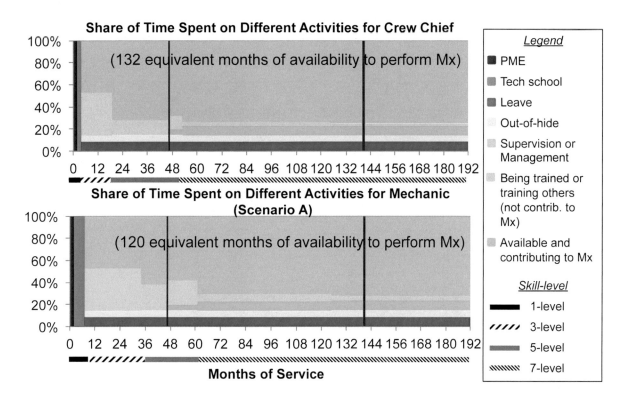

Table 2.3 summarizes the availability of the ten current AFSCs and the four consolidated AFSCs under Scenarios A and B. The average, after weighting by the authorized KC-135 personnel at the three AMC based by AFSC, falls from 132 months of availability to 120 and 125 months in Scenarios A and B, respectively. The largest reduction in availability is observed in the avionics track, which requires more technical school time than the other consolidated AFSs. It must be noted that the crew chief and mechanics track comparison is representative of the overall effect we observe. This is not surprising, given that the crew chief track is currently the largest maintenance track, comprising approximately 40 percent of the authorized KC-135 personnel in the ten AFSCs at the three AMC AFBs considered in this analysis.

**Table 2.3. Maintainer Availability Under Current and Consolidated Air Force Specialty Constructs Over the Course of a 16-Year Career**

| | | | Currently | Scenario A | | Scenario B | |
|---|---|---|---|---|---|---|---|
| | | | Mx Availability (months) | Mx Availability (months) | % Change | Mx Availability (months) | % Change |
| Avionics | 2A8X1G | CNAV | 130 | | -11 | | -7 |
| | 2A8X2G | IFCS | 131 | 116 | -11 | 121 | -7 |
| | 2A6X6 | ELEN | 131 | | -11 | | -7 |
| Mechanic | 2A5X1/4A | Crew Chief | 132 | | -9 | | -5 |
| | 2A6X1C | Propulsion | 132 | 120 | -9 | 125 | -5 |
| | 2A6X5 | Hydraulics | 132 | | -9 | | -5 |
| Structures | 2A7X1 | Metals | 132 | | -10 | | -6 |
| | 2A7X2 | NDI | 132 | 119 | -10 | 124 | -6 |
| | 2A7X3 | Structures | 132 | | -10 | | -6 |
| Fuels | 2A6X4 | Fuels | 133 | 133 | No Change | 133 | No Change |
| **Weighted Average** | | | **132** | **120** | **-9** | **125** | **-5** |

## Adjusting for Attrition

The analysis shown above does not take into account maintainer attrition. In order to account for attrition, data on the year-of-service (YOS) profile of all active-duty Air Force maintainers observed during fiscal year (FY) 2010 to 2014 are collected.[30] From this data, the retention rate profile needed to replicate the current YOS profile observed in the data is extrapolated, assuming the number of service members who joined the active-duty Air Force in maintenance occupational specialties was constant over time.

Due to issues with the assignment of personnel to AFSs, the retention-rate calculations must effectively start in the second YOS. In Figure 2.3, by year seven (84 months of service), the overall retention rate falls to approximately 50 percent; this implies that one-half of active duty maintainers separate from service by seven years. Approximately one-quarter of all active-duty maintainers remain with the Air Force for the full 16-year (192-month) period.

---

[30] In sensitivity analysis, YOS data for specific fiscal years (starting in FY 2010) was used and did not materially impact the results.

**Figure 2.3. Implied Retention Rate from Years of Service Profile
for Maintainers Observed in Fiscal Year 2010–2014**

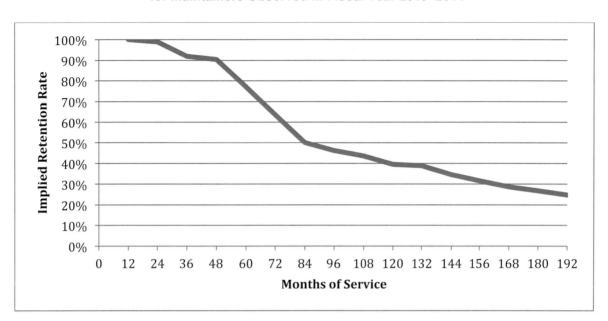

SOURCE: Extrapolated by RAND from the average YOS profile of all active-duty personnel who are associated with an AFSC that begins with "2A" in the Air Force personnel data between FY 2010 and 2014.

The calculations of maintainer availability shown in Table 2.3 are adjusted using the retention curve shown in Figure 2.3. The result of this adjustment is shown in Table 2.4. Average availability to perform maintenance activities for the ten existing AFSs falls from 132 months (in Table 2.3) to 77 months (in Table 2.4), once retention is taken into account.

When comparing the availability estimates that take into account retention before and after AFS consolidation, one sees larger percent changes than when retention is not accounted for. For example, for a crew chief that becomes a mechanic, availability drops by approximately 14 percent in Scenario A when retention is taken into account. If the same comparison is performed over a full 16-year career for the same AFS consolidation example, the reduction in availability is approximately 9 percent in Table 2.3. The percentage reduction in availability from AFS consolidation is worse when retention is accounted for because training is frontloaded (i.e., maintainers are less productive earlier in their career).

## Table 2.4. Maintainer Availability Under Current and Consolidated Air Force Specialty Constructs After Taking into Account Retention

| | | | Currently | Scenario A | | Scenario B | |
|---|---|---|---|---|---|---|---|
| | | | Mx Availability (months) | Mx Availability (months) | Percentage Change | Mx Availability (months) | Percentage Change |
| Avionics | 2A8X1G | CNAV | 76 | | -17 | | -8 |
| | 2A8X2G | IFCS | 76 | 63 | -17 | 70 | -8 |
| | 2A6X6 | ELEN | 76 | | -17 | | -8 |
| Mechanic | 2A5X1/4A | Crew Chief | 77 | | -14 | | -7 |
| | 2A6X1C | Propulsion | 77 | 66 | -15 | 71 | -7 |
| | 2A6X5 | Hydraulics | 77 | | -15 | | -7 |
| Structures | 2A7X1 | Metals | 77 | | -16 | | -7 |
| | 2A7X2 | NDI | 77 | 64 | -16 | 71 | -8 |
| | 2A7X3 | Structures | 77 | | -16 | | -7 |
| Fuels | 2A6X4 | Fuels | 77 | 77 | No Change | 77 | No Change |
| **Weighted Average** | | | **77** | **66** | **-14** | **71** | **-7** |

16

# 3. Modeling the Impact of Air Force Specialty Consolidation on Maintenance Manpower Effectiveness

Having characterized the effect on maintainer availability due to increased training required for AFS consolidation, this chapter describes the overall impact of consolidation on maintenance manpower requirements and readiness. Gotz and Stanton[31] note that consolidation of AFSs has the potential to increase unit efficiency because personnel will sit idle during lulls in the demand for their specialties less frequently.[32] This analysis examines this potential by using LCOM and the official KC-135 LCOM data base as reported in the KC-135 LCOM Final Report,[33] the Simulated Maintenance and Requirements Tool (SMART) sheets for MacDill, McConnell, and Fairchild AFB, and data on authorized and assigned personnel levels as of the end of FY 2013.

LCOM allows the user to specify information on sortie demand, the maintenance workforce, and aircraft maintenance actions that can occur either on a scheduled or stochastic basis according to component failure rates. In LCOM, aircraft require maintenance as they are flown and maintenance crews are called on to complete inspections and repairs. The amount of time aircraft are not mission capable depends in part on the availability and productivity of maintainers. The model tracks how many scheduled sorties are actually flown over a simulated year.

For the purposes of evaluating sortie generation rates, a stressing sortie schedule is specified in LCOM and the number of requested sorties that can actually be performed under the existing and consolidated AFS constructs is recorded. Within LCOM, AFS consolidation has the potential to improve sortie generation rates by increasing the number of maintainers able to work on certain repairs or inspections. That is, aircraft are less likely to be non-mission capable due to maintainer shortages when maintainers are trained to perform a broader range of tasks.[34] As a result, under AFS consolidation, there is the potential to maintain the same sortie generation rate

---

[31] Glenn A. Gotz and Richard E. Stanton, *Modeling the Contribution of Maintenance Manpower to Readiness and Sustainability*, Santa Monica, Calif.: RAND Corporation, R-3200-FMP, 1986.

[32] While not considered here, Layne et al., 2001, note that "[t]here is also the possibility of increased quality and adaptability as the work within units becomes more process oriented. Personnel with broader skills better understand how different tasks blend into a total desired output and feel more empowered to make meaningful contributions."

[33] Shenita Clay, *KC-135 Logistics Composite Model (LCOM) Final Report: Peace and Wartime—Peacetime Update*, Scott Air Force Base, Ill.: HQ AMC/XPMMS, May 1, 1999.

[34] Some subject-matter experts voiced concern that broadening maintainer task responsibilities could reduce maintainer proficiency or increase the amount of time required to conduct certain maintenance tasks. However, others suggested that expanding training and increasing maintainer utilization could increase proficiency and improve maintenance times. This study assumes that consolidating AFSs will not positively or negatively affect the proficiency and speeds with which maintenance tasks are performed. This is an area where future research would be valuable.

with fewer maintainers or achieve greater sortie generation rates if the workforce size is not changed. Offsetting this benefit is the reduced maintainer availability analyzed in the previous chapter. In this chapter, the improved use of maintainers and their reduced availability after AFS consolidation are both integrated into the analysis.

LCOM simulates activity at the squadron or wing level and is used to determine home stationed and deployed manpower requirements.[35] The SMART sheet for each base combines the home station and deployed manpower requirements derived from LCOM runs and makes other adjustments due to shift and leave policies and other factors to determine the overall maintenance manpower requirement. The KC-135 manpower requirements are largely driven by home-station flying requirements.[36] As a result, attention is focused on modeling home-station requirements; wartime impacts are assumed to be proportional.

It is important to note that LCOM determines a portion, but not all, of the KC-135 maintenance manpower requirement at McDill, McConnell, and Fairchild AFBs. Based on analysis of the SMART sheets for those AFBs, the positions that compose maintenance manpower requirements are classified into three groups, as follows:

- Group 1 (in blue in Figure 3.1) makes up 72 percent of the maintenance manpower requirement and is based on LCOM. This requirement covers the ten AFSCs considered in this analysis.
- Group 2 (in yellow in Figure 3.1) makes up 22 percent of the manpower requirement and is associated with the ten AFSCs considered in the analysis, but this requirement is not determined using LCOM. The majority of these positions represent maintenance expeditors and production superintendent positions.
- Group 3 (in red in Figure 3.1) makes up 6 percent of the maintenance manpower requirement and is associated with maintenance AFSCs that start with "2A," but are not one of the ten "2A" AFSCs considered in the analysis. The largest AFSC in this group is 2A6X2, aerospace ground equipment.

Each group is treated differently for the purpose of determining the overall effect of AFS consolidation on manpower requirements, as described in Figure 3.1. Specifically, for Group 1, AFS consolidation reduces maintainer availability (as described earlier in this report), but improves maintainer capability according to the analysis conducted using LCOM. For Group 2, AFS consolidation reduces maintainer availability but is assumed to have no effect on maintainer capability (since these maintainers are presumed to be engaged in management or other activities that do not benefit from the additional training required under AFS consolidation). This is likely a conservative assumption. Finally, Group 3 is assumed to be unaffected by AFS consolidation,

---

[35] For a detailed description of how LCOM is used to determine maintenance manpower requirements, see Dahlman, Kerchner, and Thaler, 2002.

[36] Specially, it was found that the vast majority of the manpower requirements determined using LCOM were derived from peacetime, rather than wartime, scenarios when reviewing the KC-135 SMART sheets for the three AMC bases considered in this analysis.

since these maintainers belong to AFSCs outside the ten considered. These assumptions are applied to derive the impacts of AFS consolidation on the maintainer base as a whole.

**Figure 3.1. Treatment of Different Maintenance Groups in Impact Calculations**

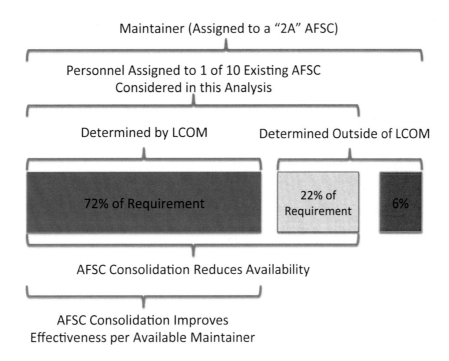

Currently, McDill, Fairchild, and McConnell AFB have 16, 30, and 44 KC-135 primary aircraft authorizations (PAA), respectively. PAA and aircraft use rates drive maintenance manpower requirements and can be varied in the SMART sheet for each AMC KC-135 AFB. LCOM runs for a 16 and 48 PAA KC-135 wing are developed and used to extrapolate findings for the KC-135 wings operated out of McDill, McConnell, and Fairchild AFBs.

## Impact of Air Force Specialty Consolidation on Manpower Requirements and Readiness

The effect of consolidating AFSCs on sortie generation rates under varying maintenance manpower levels is shown in Figure 3.2.[37] In the figure, the black dashed, green, and blue curves

---

[37] To derive Figure 3.2, we first generate LCOM runs that assume the existing AFS structure and that vary the manning level under the AFS consolidation structure described in Figure 1.1. We next expand the LCOM manning levels to represent total maintenance manpower requirements consistent with the methodology employed in the KC-135 SMART. The LCOM runs cover approximately 72 percent of the total maintenance manpower requirement, as described in Figure 3.1. The results of these steps allow us to derive the black dashed curve relative to the red dot in Figure 3.2. To generate the green and blue curves, we increase the manning levels underlying the black dashed curve to account for the reduced maintainer availability of 7 and 14 percent from AFS consolidation quantified in Table 2.4. The reduced maintainer availability from AFS consolidation effects 94 percent of maintainers as described in Figure 3.1.

represent the effect of AFS consolidation at different authorized manning levels on sortie generation capability after the effects outlined in Figure 3.1 are accounted for, and assuming reductions in maintainer availability of 0 percent, 7 percent (consistent with Scenario B), and

**Figure 3.2. Impact of Consolidating Maintenance Air Force Specialty Codes on Sortie Generation Capability and Manpower Requirements**

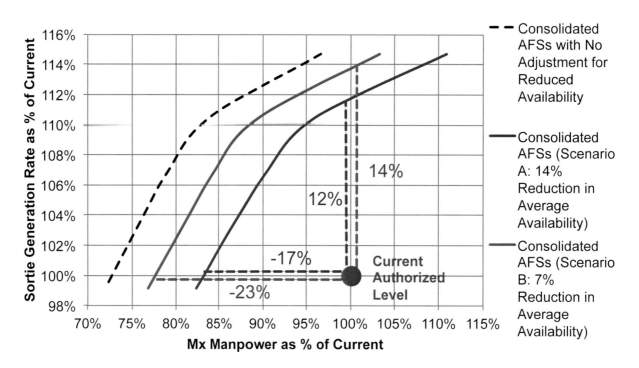

14 percent (consistent with Scenario A), respectively. These curves should be interpreted relative to the red dot, which represents the current manning authorization level and the associated sortie generation rate.

Figure 3.2 suggests that consolidating KC-135 maintenance AFSs according to the construct analyzed here would, in the long run, benefit AMC through greater readiness (as measured by sortie generation capability), reduced manpower requirements, or both. If current manpower authorization levels are maintained and fully staffed, KC-135 sortie generation capability would increase by between 12 and 14 percent under the AFS consolidation construct analyzed. Alternatively, AMC could reduce its KC-135 maintenance manpower authorization levels by between 17 and 23 percent, without experiencing any loss in sortie generation capability. A third option involves reducing manpower levels by less than 17 percent and reaping some reduction in manpower costs, while also improving sortie generation capability.

## Authorized Versus Assigned Personnel Levels

The findings presented in Figure 3.2 correspond to authorized personnel levels and do not consider the fact that the Air Force tends not to fill all authorized manpower positions. The study team was also asked to consider the implications of AFS consolidation on assigned (rather than authorized) personnel levels. To do this, the framework shown above was adapted.

Data on the number of active-duty KC-135 maintainers authorized and assigned at MacDill, McConnell, and Fairchild AFB suggests that, for every 100 authorized positions, there are only 91 assigned personnel. That is, the so-called fill rate varies by base and across AFSs, but it averaged 91 percent at the end of FY 2013 for active-duty "2A" positions at the three bases we considered.

Figure 3.3 extends the curves shown in Figure 3.2 to the left and adds a curve illustrating how sortie generation capability falls under the existing AFSC construct when manpower is reduced in the red dotted line. The purple dot in Figure 3.3 represents the readiness level associated with the current assigned personnel level (which is 91 percent of the authorized level). According to the chart, AMC's KC-135 fleet would have approximately 4 percent greater sortie generation capability if it manned at 100 percent of the authorized level.

After AFS consolidation, the assigned personnel levels could be reduced by between 11 to 17 percent without degrading readiness below the level associated with the current assigned manning level, or it could maintain the current number of assigned personnel and improve sortie generation capability by between 11 and 16 percent. These findings further support the long-run benefits of consolidating AFSs, although the exact impacts vary somewhat from those shown in Figure 3.2. For example, the manpower reductions that can be achieved without loss of readiness drop by approximately 5 percent when assigned rather than authorized personnel levels are considered.

**Figure 3.3. Impact of Consolidating Maintenance Air Force Specialties on Sortie Generation Capability at Varying Assigned Personnel Levels**

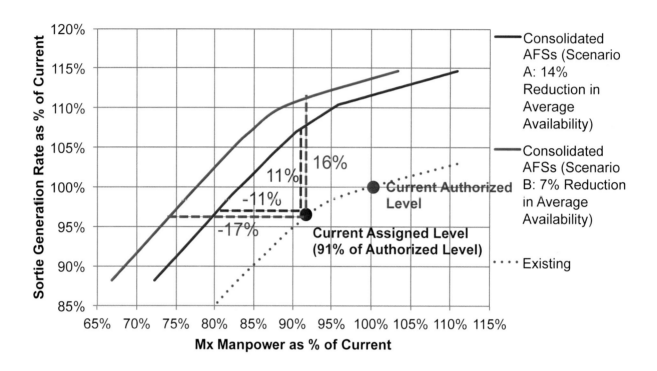

# 4. Long-Run Cost and Readiness Implications of Air Force Specialty Consolidation

In the previous chapter, it was shown that AFS consolidation has the potential to reduce manpower requirements and improve readiness. This chapter explores the long-run monetary implications of AFS consolidation at different manning levels. The calculation of the monetary impacts includes the reduction in spending on authorized personnel as well as the increased cost of training that would be absorbed by AETC.

## Cost Savings Obtained at Different Levels of Readiness Improvement

Table 4.1 illustrates how consolidating AFSs and varying the size of the manpower authorization cuts perform in terms of sortie generation rates and cost savings. The cost-saving analysis assumes that each eliminated active-duty position would save $87,600 per year in FY 2014 dollars.[38] For example, if a 15 percent manpower cut is pursued following AFS consolidation, this would result in 380 fewer active-duty KC-135 maintainers positions at McDill, Fairchild, and McConnell AFB, which translates into a saving of $33.3 million per year in FY 2014 dollars (= 380 fewer authorizations x $87,600 saved per authorization). This reduction in authorized personnel levels is associated with a 2–8 percent improvement in sortie generation capability if it is implemented after consolidating AFSs.

If the Air Force pursues AFS consolidation for the KC-135 and changes its manpower authorization levels, the training burden placed on AETC would change, further affecting Air Force costs. In order to estimate the magnitude of the impact on AETC training costs, data on the cost of technical school training for existing KC-135 maintenance AFSCs is analyzed, and the cost of extending technical school classroom time in line with those outlined earlier in this report for Scenarios A and B is extrapolated. This analysis suggests that, on average, the cost to AETC per student will increase by between $13,000 and $7,000 per recruit in training Scenarios A and B (from a base of approximately $25,000 per recruit), respectively.[39]

In the calculations involving manpower reductions, the increase in AETC training costs from AFS consolidation is offset partially by the fact that fewer total recruits must be trained each

---

[38] Secretary of the Air Force, "Military Composite Pay," Table A19-2, updated c. 2013, for *U.S. Air Force Cost and Planning Factors*, Air Force Instruction 65-503, February 4, 1994. Table A19-2 referenced by author on October 2, 2013, and has since been revised.

[39] The effect on AETC is calculated for each existing and consolidated AFS and combined using the manning level in each AFSC to derive the averages reported here. The cost estimates exclude the cost of compensating and recruiting each new recruit, which AETC includes in its cost estimates. These costs are not included because they will be borne regardless of whether the recruit is attending technical school or starting his or her assignment at their first base.

year. In the case where there are no manpower reductions, training costs increase by between $1.7 and $3.2 million per year.[40] But, as shown in Table 4.1, the additional AETC training cost is smaller in cases where a cut to the workforce is taken because fewer individuals must be trained each year to maintain a stable workforce size.

**Table 4.1. Cost Impacts of Adopting Air Force Specialty Consolidation and Varying Manpower Levels**

| Percentage of Reduction in Mx Manpower Authorizations After AFS Consolidation | Percentage of Sortie Generation Rate Relative to Current | Mx Manpower Authorization Reduction at AMC KC-135 Bases[a] | Millions of FY 2014 Dollars per Year | | |
|---|---|---|---|---|---|
| | | | Mx Manpower Savings[b] | Increased AETC Training Costs[c] | Net Cost Reduction |
| None | 112–114 | None | None | $1.7–$3.2 | -$1.7 to -$3.2 |
| 5 | 110–112 | 127 | $11.1 | $1.3–$2.7 | $8.4–$9.8 |
| 10 | 107–111 | 253 | $22.2 | $0.9–$2.2 | $20.0–$21.3 |
| 15 | 102–108 | 380 | $33.3 | $0.5–$1.8 | $31.5–$32.8 |

[a] Reduction is from a base on 2,531 funded and authorized active duty 2As at MacDill, McConnell, and Fairchild AFBs associated with KC-135.
[b] Assumes savings per manpower position of $87,600 in FY 2014 dollars (Air Force Instruction 65-503, June 10, 2013).
[c] Impact on AETC training costs estimated by RAND.

The last column in Table 4.1 shows the net effect of offsetting any savings associated with fewer authorized personnel against the increased cost to AETC. In the case where no personnel cuts are made, the costs are estimated to increase by $1.7 to $3.2 million per year. If a 15 percent personnel cut is made, costs in the long run would be $31.5–$32.8 million less per year. A range of options exists between these two endpoints, as illustrated in Table 4.1. Note that all the cases considered in Table 4.1 improve the readiness metric (e.g., the sortie generation rate under a stressing sortie demand schedule).

## Caveats and Next Steps

This analysis suggests that AFS consolidation has the potential in the long run to reduce manpower costs and/or improve readiness for the KC-135 fleet. Some caveats to this analysis

---

[40] This calculation assumes that 246 new active-duty personnel must join the force and be trained as KC-135 maintainers to maintain a workforce of 2,531 active-duty 2As at MacDill, McConnell, and Fairchild AFBs associated with the KC-135. This "replacement rate" of approximately one-tenth the workforce (= 246/2531) is consistent with the rate required, assuming the retention curve analysis presented earlier. Assuming an additional training cost of $7,000–$13,000 per recruit, this leads us to a cost range of $1.7 (= 246 recruits x $7,000 per recruit) to $3.2 million (= 246 recruits x $13,000) per year in additional training costs.

should be considered, however, to better inform decisionmakers on the range of impacts they should expect following AFS consolidation.

First, this analysis assumes that a reduction in the number of maintenance AFSs would be accompanied by significant increases in training so that maintainers would have the same capability (depth of knowledge and skill) to perform maintenance actions as today's maintainers, but that capability would be spread over a broader range of tasks for each maintainer. Should either the training time requirements or the assumptions on the depth of maintainer knowledge from training not hold, the findings of this analysis would be different. Subject-matter experts voiced particular concern with these assumptions, since past Air Force experience with AFS consolidation required that training be expanded beyond what was initially assumed by the Air Force or that AFS consolidation needed to be reversed due to concerns associated with excessive generalization of maintenance skills.[41]

Second, a number of Air Force officials the study team spoke with raised concerns stemming from the fact that the Air Force has historically under-filled authorized maintenance positions, and that any initiative that could lead to a reduction in manpower levels should formally address this. The study team recommends that policies to reduce manpower authorizations should be accompanied by steps that improve fill rates and should be implemented gradually for legacy fleets so that actual experience with consolidation impacts can inform any manpower cuts that are made.

Third, this analysis only considers a subset of the costs and benefits of AFS consolidation, and some aspects of this policy decision would benefit from additional analysis. In particular, the study team recommends that the Air Force:

- **Investigate the impact on Guard and Reserve maintainers.** This analysis focused on the impact of AFS consolidation on active-duty personnel at AMC bases. Additional research is needed to understand how implementation of fewer AFSs will affect Guard and Reserve maintainers, as well as operations at non-AMC bases.
- **Identify implementation challenges and develop strategies to address them.** Past efforts to reduce the number of maintenance AFCs, such as the Rivet Workforce initiative implemented in the 1980s and 1990s, have disrupted operations and required costly and unanticipated adjustments according to subject-matter experts. Layne et al. (2001) notes that changes to military occupational structure tend to occur over 24-month periods, over which disruptions and turbulence can occur.
  - Before AFS consolidation is adopted, effort should be made to identify potential implementation challenges and develop strategies for reducing those issues. Lessons learned from the Rivet Workforce initiative and the more recent effort to reduce the number of maintenance AFSs for fifth-generation fighters (the F-22 and F-35) could

---

[41] This concern was raised in the context of the Rivet Workforce initiative and more recently in the context of eliminated "shreds" for certain AFSCs. In these cases, subject matter experts indicated that affected maintainers were not necessarily adequately trained to perform the broader set of maintenance responsibilities assigned to them, following consolidation.

prove valuable. Although implementation differences for legacy and new fleets was not formally analyzed in this study, the study team anticipates that the turbulence caused by a transition to fewer AFSs will be less for new fleets such as the KC-46. As a result, this concept should first be considered in the context of the KC-46 and later for legacy MAF fleets.

- **Evaluate the effects on other fleets.** This research considered the effect of AFS consolidation on AMC's KC-135 fleet and was intended to inform KC-46 maintenance decisions that AMC is currently developing. The analysis could be replicated for the C-130, C-17, C-5, and KC-10 fleets to better identify the broad applicability of maintenance AFS consolidation for the MAF.

- **Refine estimates of training requirements associated with technical school and upgrade and qualification training.** This analysis expands training requirements under AFS consolidation based on input from subject-matter experts and using simple analysis. As shown in Chapter Two, the assumptions made in Scenarios A and B have a significant effect on training requirement and the estimates of maintainer availability. Future analysis could refine the training requirements assumptions used in this analysis by working with AETC and the Major Commands to more fully consider technical school course curricula and training requirements contained in the CFETP for new AFSs.

- **Evaluate the impacts on accession and promotion requirements.** This analysis has not considered the effects of AFS consolidation on accession and promotion requirements. Currently, accession requirements are different for some of the AFSs considered for consolidation, and this will need to be worked out. This could impact the relative attractiveness of certain AFSs to recruits, which could have net positive or negative implications for recruiting requirements and costs. Combining AFSs will have promotion implications as well. With additional tasks included in an AFS, it will potentially make the skill test in the Specialty Knowledge Test more difficult, impacting career progression.

- **Evaluate impacts on deployed footprint and combat resiliency.** A consolidated maintenance construct could reduce manpower and resource deployment requirements. Depending on the number of maintainers deployed, *combat resiliency* (e.g., the ability to launch aircraft following an attack at a forward operating location) could also be enhanced or reduced.[42] Future research should address these issues for both Combat Air Force and MAF fleets.

Finally, it is worth noting that other initiatives being considered by AMC could work synergistically with AFS consolidation, potentially enhancing the size of the net benefits shown here. These include expanding cross-utilization training prior to AFS consolidation, staggering technical school training by skill level, combining AETC's Field Training Detachment and AMC's MQTP, and consolidating common tasks in CFETP when combining AFSs.

---

[42] If the number of maintainers deployed is not changed after AFS consolidation, combat resiliency is expected to improve. However, if AFS consolidation is accompanied with fewer deployed maintainers, it is unclear how combat resiliency would be impacted.

# Appendix A: Comparison of United Airlines and Mobility Air Force Maintenance Practices

This appendix provides a summary of differences in maintenance practices employed by United Airlines (UA) at its George Bush Intercontinental Airport operations in Houston, Texas, and at its Technical Operational Maintenance Control (TOMC) center in Chicago, Illinois, in addition to those used by the Air Force to maintain mobility aircraft.

Figure A.1 illustrates some of the key organizational features of UA's and the MAF's maintenance structure.[43] Starting with personnel differences, the majority of current UA hires hold a general Airframe and Powerplant (A&P) certification. An A&P certification is the basic Federal Aviation Administration (FAA)–approved rating that certifies that the holder has basic aviation knowledge and skills required to maintain commercial aircraft. To receive an A&P certificate, one receives training in:

- general aircraft maintenance activities (crew chief tasks)
- electricity and power plant
- the content and structure of technical documentation
- most aircraft subsystems (e.g., airframe, hydraulics, fuels, environmental, electrical).

The technical requirements for A&P certification are outlined in FAA's publication in 2008.[44]

At UA, each new hire is supervised in their first two years. After that point, a maintainer at UA is expected to have become certified as an Aircraft Airworthiness Release Authority (AARA) on at least one type of aircraft. An AARA is responsible for ensuring maintenance is performed according to manuals and no known condition exists that render the aircraft unfit for flight. An AARA can certify an aircraft is airworthy after maintenance has been performed.

After receiving AARA on his or her first platform, a UA maintainer continues to obtain AARA on other aircraft platforms (e.g., mission design series [MDS], using Air Force terminology). After obtaining the appropriate experience, training (formal and on-the-job), and supervisor recommendations, the maintainer can also begin specializing in one of the following three maintenance specialty areas (or remain an A&P generalist):

---

[43] Figure A.1 is not intended to show chain of command. For example, UA's line and base maintenance organizations do not report to the TOMC center for supervisory management purposes; and MXGs do not report to the Tanker Airlift Control Center (TACC) or depots. They are intended to show the organizations from which maintenance information may come.

[44] Federal Aviation Administration, *Aviation Maintenance Technician—General, Airframe, and Powerplant Knowledge Test Guide*, FAA-G-8082-3A, September 2008.

- *Avionics:* The avionics specialty at UA is able to conduct repairs and maintenance similar to AFSs in Communications and Navigation (CNAV), Instrument and Flight Control Systems (IFCS), and some Electro-Environmental (ELEN) systems.
- *Structures:* The structures specialty is most similar to structural maintenance, metals technology, and composite system maintenance AFSs.
- *Interior:* The interior specialty is most similar to fabrication maintenance (e.g, seats, flooring, cabin lining) and some life-support equipment maintenance (e.g., life rafts and emergency escape slides).

**Figure A.1. Comparison of United Airlines and Air Force Maintenance Structures**

*United Airlines Maintenance Structure*

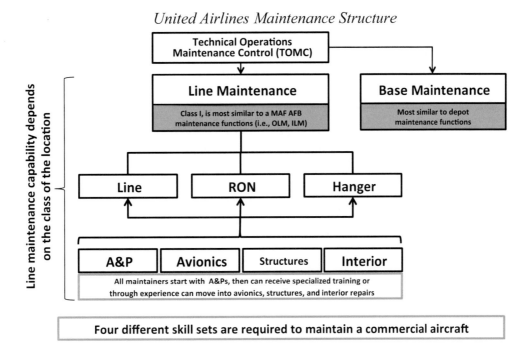

Four different skill sets are required to maintain a commercial aircraft

*Air Force Maintenance Structure for Mobility Aircraft*

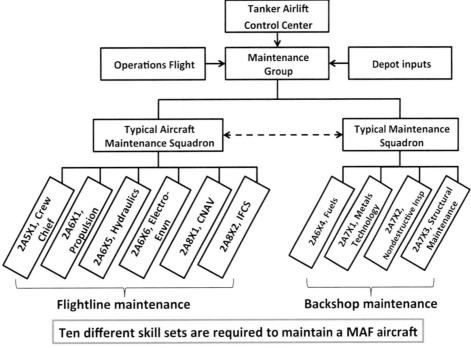

Ten different skill sets are required to maintain a MAF aircraft

The MAF maintenance structure trains maintainers into one of ten AFSs, where they remain under most circumstances. MAF maintenance specialists receive some cross-utilization training after technical school training, once they arrive at their first duty station through a mandatory

29

MQTP.[45] It should be noted that many general maintenance activities—such as aircraft marshalling, towing, fueling, and deicing/anti-icing—are not performed by UA maintainers referred to in this research, whereas MAF maintainers do perform these tasks.

As shown in Figure A.1, UA divides its maintenance organization into two broad categories, *line maintenance* and *base maintenance*. Line maintenance can be compared with maintenance activities performed on an operational MAF base. UA's line-maintenance function is divided into three sections:

- Line maintenance personnel are responsible for maintenance of aircraft in the daily flying schedule. Each UA concourse is divided into zones with assigned maintainers who are responsible for maintaining any MDS that arrives.
- Remain overnight personnel maintain all MDSs as needed to prepare for the next day's flying schedule. This is primarily night shift work focusing on aircraft in the next day schedule.
- Hanger maintenance personnel perform scheduled and extended unscheduled maintenance. Aircraft that are scheduled for a "Line Special Visit" are kept down for a certain number of hours to perform, for example, extended phase/isochronal inspections, modifications, or time-compliance maintenance.

Together, these three sections perform tasks that are equivalent to MAF flight line maintenance activities (i.e., operational-level maintenance, sortie generation) and backshop maintenance activities (i.e., intermediate-level maintenance, off-equipment).

Under UA's structure, a number of maintainers from each specialty are employed in each line maintenance section to maintain all aircraft. Maintainers can move from section to section for further broadening as union rules dictate. Under the MAF structure, the ten maintainer AFSs are assigned typically into AMX (flight line) and MX (backshop) squadrons.[46]

The level of capability at each location varies depending on the level of UA operations conducted. UA identifies each location based on its class:

- Class I—Has 24/7 full-service line maintenance capability. A Class I location is expected to be able to maintain all arrival aircraft, perform any level of line maintenance, including extensive unscheduled maintenance, such as fuel-tank repair, engine repair/replacement, and structural repairs; and extensive scheduled maintenance, including letter check inspections.

---

[45] Level 1 of a MAF specialist's MQTP consists of 19 academic days to learn general maintenance (e.g., safety topics, forms and technical data use), production-team maintenance tasks (e.g., operating power sources, surfaces, and doors; assist in refueling and aircraft movements), duty position tasks (Headquarters AMC/A4M, KC135-2AXXXX-1, MQTP Training Plan, Course Chart and Training Standard KC-135 General and Production Team Maintenance Training, June 26, 2013).

[46] It is understood that some AFSs shown as primarily employed in flight line maintenance can also perform backshop maintenance activities, and that some AFSs that are shown as assigned to the backshop perform unscheduled flight line maintenance.

- Class II—Has 24/7 service, but is less than Class I line maintenance capability. A Class II location may not be able to maintain all arrival MDSs if in-depth maintenance is required, or not capable of performing hangar-level maintenance.
- Class III—Has a more limited line-maintenance capability. It may share maintenance activities with other onsite carriers.
- Class IV—Has an on-call contractor-support agreement.

The UA *base maintenance* function can be likened to an Air Force depot operation (e.g., in-depth inspections, Line Replaceable Unit repair, engine overhaul) and is performed by a completely separate organization than line and was not analyzed any further in this research. Base maintenance operations are located at some Class I UA locations.

UA's class system is similar to the MAF's global en-route tiered maintenance support system (e.g., tier one, major maintenance, full service; tier two, minor maintenance, in-transit capability; tier three, limited maintenance and expeditionary [as mission dictates]). Within the contiguous United States main operating base (MOB) structure, each MOB has all required support functions to maintain assigned MDSs and is similar in maintenance capability to a UA Class I airport. However, regardless of the class of a UA location, no MDSs are specifically assigned to any one location and continuously transit their network. The MAF has a more restrictive network of maintenance capabilities capable of working on specific MDSs.

The TOMC center is part of UA's Network Operation Center (NOC),[47] an enterprise-level organization that controls day-to-day worldwide operations. The TOMC center is an aircraft technical-support group in the NOC that plans for and schedules maintenance activities while meeting UA's flight schedule and communicating with flight crew and maintainers worldwide. It uses sophisticated information technology to monitor and plan both the operations and maintenance of aircraft. Some specific functions of the TOMC include:

- provide authority for all deferrals and monitor all aircraft through daily and weekly aircraft reviews, including generating work directives through callouts, for the overall maintenance program during the execution of maintenance activities
- provide flight crews with technical assistance or coordination when necessary to assist with the operational control of the aircraft during flight
- plan and coordinate repairs to aircraft at non-maintenance stations
- coordinate the planning and repair of chronic aircraft defects
- provide deferred maintenance quality control
- balance aircraft operating system workload
- coordinate the implementation of various line-maintenance processes
- marshal maintenance and engineering resources to support the daily operation
- execution of applicable maintenance and engineering crisis policies and procedures
- plan for bringing together parts/maintenance.

---

[47] Some other functions of the NOC include flight planning, weight and balance, flight following, aircraft routing, crew management, and weather tracking.

Some of the functions performed at the NOC resemble AMC's Tanker Airlift Control Center (TACC) operations. Other functions performed at the NOC resemble functions performed at each MAF base support organizations, such as Maintenance Operations Flight's maintenance operations control centers, plans and scheduling sections, Air Force Engineering and Technical Services, and depot's engineering support.

Table A.1 summarizes some of the main differences we observed between MAF and UA maintenance structures. A key difference is the maintenance occupational structures observed at UA and in the MAF. This report explores the implications of a MAF maintenance occupational structure that is more consolidated and similar (although not identical) to that employed by UA.

**Table A.1. Summary of Mobility Air Force and United Airlines Maintenance Structure Differences**

| United Airline Approach | USAF (MAF) Approach |
|---|---|
| • Hires basic technically trained personnel (A&P) | • Hires unqualified personnel, conducts all training |
| • A&P maintainers may specialize into one of three areas or remain as A&P generalists | • Maintainers specialize into one of ten AFSs |
| • Crew chief tasks are performed by every maintainer | • Has a separate crew chief specialty |
| • Certain ground-handling operations (e.g., marshalling, tow, deicing, fueling) performed by different pool of personnel | • Maintainers perform/involved in all aircraft ground handling |
| • Maintainers maintain multiple MDSs concurrently | • Maintainers maintain one MDS at a time |
| • Remain at same location for extended periods | • PCS periodically |
| • Use sophisticated information technology systems that link operations and Mx | • Limited information sharing between operations and Mx functions |
| • Aircraft are not assigned to any one location | • Aircraft are assigned to each MAF base |

# References

Albrecht, Mark J., *Labor Substitution in the Military Environment: Implications for Enlisted Force Management*, Santa Monica, Calif.: RAND Corporation, R-2330-MRAL, 1979. As of November 3, 2015:
http://www.rand.org/pubs/reports/R2330.html

Boyle, Edward, Stanley J. Goralski, and Michael D. Meyer, "The Aircraft Maintenance Workforce Now and in the Twenty-First Century," *Air Force Journal of Logistics*, fall 1985, pp. 3–5.

Clay, Shenita, *KC-135 Logistics Composite Model (LCOM) Final Report: Peace and Wartime— Peacetime Update*, Scott Air Force Base, Ill.: HQ AMC/XPMMS, May 1, 1999.

Da Cunha, Mozer O., "NCOA Changes Curriculum and Eligibility Requirements," *Inside Barksdale AFB*, November 4, 2014, updated November 5, 2014. As of August 1, 2015:
http://www.barksdale.af.mil/news/story.asp?id=123430502

Dahlman, Carl J., Robert Kerchner, and David E. Thaler, *Setting Requirements for Maintenance Manpower in the U.S. Air Force*, Santa Monica, Calif.: RAND Corporation, MR-1436-AF, 2002. As of November 3, 2015:
http://www.rand.org/pubs/monograph_reports/MR1436.html

Drew, John, Kristin F. Lynch, James M. Masters, Robert S. Tripp, and Charles Robert Roll, Jr., *Options for Meeting the Maintenance Demands of Active Associate Flying Units*, Santa Monica, Calif.: RAND Corporation, MG-611-AF, 2008. As of November 3, 2015:
http://www.rand.org/pubs/monographs/MG611.html

Federal Aviation Administration, *Aviation Maintenance Technician—General, Airframe, and Powerplant Knowledge Test Guide*, FAA-G-8082-3A, September 2008. As of November 3, 2015:
https://www.faa.gov/training_testing/testing/test_guides/media/faa-g-8082-3A.pdf

Headquarters AMC/A4M, KC135-2AXXXX-1, MQTP Training Plan, Course Chart and Training Standard KC-135 General and Production Team Maintenance Training, June 26, 2013.

Gotz, Glenn A., and Richard E. Stanton, *Modeling the Contribution of Maintenance Manpower to Readiness and Sustainability*, Santa Monica, Calif.: RAND Corporation, R-3200-FMP, 1986. As of November 3, 2015:
http://www.rand.org/pubs/reports/R3200.html

Layne, Mary E., Scott Naftel, Harry J. Thie, and Jennifer H. Kawata, *Military Occupational Specialties: Change and Consolidation*, Santa Monica, Calif.: RAND Corporation, MR-977-OSD, 2001. As of November 3, 2015:
http://www.rand.org/pubs/monograph_reports/MR977.html

McGarvey, Ronald G., Manuel Carrillo, Douglas C. Cato, John G. Drew, Thomas Lang, Kristin F. Lynch, Amy L. Maletic, Hugh G. Massey, James M. Masters, Raymond A. Pyles, Ricardo Sanchez, Jerry M. Sollinger, Brent Thomas, Robert S. Tripp, and Ben D. Van Roo, *Analysis of the Air Force Logistics Enterprise: Evaluation of Global Repair Network Options for Supporting the F-16 and KC-135*, Santa Monica, Calif.: RAND Corporation, MG-872-AF, 2009. As of November 3, 2015:
http://www.rand.org/pubs/monographs/MG872.html

McGarvey, Ronald G., James H. Bigelow, Gary J. Briggs, Peter Buryk, Raymond E. Conley, John G. Drew, Perry Shameem Firoz, Julie Kim, Lance Menthe, S. Craig Moore, William W. Taylor, and William A. Williams, *Assessment of Beddown Alternatives for the F-35*, Santa Monica, Calif.: RAND Corporation, RR-124-AF, 2013. As of November 3, 2015:
http://www.rand.org/pubs/research_reports/RR124.html

McMahan, Matthew, *Maintenance Force Development Study: Phase II Final Report*, Booz Allen Hamilton, April 2, 2007.

———, *Maintenance Force Development Study: Final Report*, Booz Allen Hamilton, June 17, 2008.

Oliver, Steven A., *Cost and Valuation of Air Force Aircraft Maintenance Personnel Study*, Maxwell AFB, Gunter Annex, Ala: Air Force Logistics Management Agency, August 2001.

Secretary of the Air Force, "Military Composite Pay," Table A19-2, updated c. 2013, for *U.S. Air Force Cost and Planning Factors*, Air Force Instruction 65-503, February 4, 1994. Table A19-2 referenced by author on October 2, 2013, and has since been revised.

Tripp, Robert S., Ronald G. McGarvey, Ben D. Van Roo, James M. Masters and Jerry M. Sollinger. *A Repair Network Concept for Air Force Maintenance: Conclusions from Analysis of C-130, F-16, and KC-135 Fleets*, Santa Monica, Calif.: RAND Corporation, MG-919-AF, 2010. As of November 3, 2015:
http://www.rand.org/pubs/monographs/MG919.html

Van Roo, Ben D., Manuel Carrillo, John G. Drew, Thomas Lang, Amy L. Maletic, Hugh G. Massey, James M. Masters, Ronald G. McGarvey, Jerry M. Sollinger, Brent Thomas, and Robert S. Tripp, *Analysis of the Air Force Logistics Enterprise: Evaluation of Global Repair Network Options for Supporting the C-130*, Santa Monica, Calif.: RAND Corporation, TR-813-AF, 2011. As of November 3, 2015:
http://www.rand.org/pubs/technical_reports/TR813.html

Wild, William G., Bruce R. Orvis, Rebecca Mazel, Iva Maclennan, and R. D. Bender, *Design of Field-Based Crosstraining Programs and Implications for Readiness*, Santa Monica, Calif.: RAND Corporation, R-4242-A, 1993. As of November 3, 2015: http://www.rand.org/pubs/reports/R4242.html